熊澄宇　主编
Editor in Chief　Xiong Chengyu

文化藝術出版社
Culture and Art Publishing House

TAI RESIDENCE

A REALM WHEREIN THE TRUE SELF LIES

九层台 观自在

目 录
Contents

卷首语

PROLEGOMENON

“合抱之木，生于毫末；九层之台，起于累土。”这既是过程，又是结果；是现实，又是理想。由种子而参天，历经沧桑；聚细沙而成塔，贵在坚持。登九层高台，澄怀观道。

Laozi, one of the greatest philosophers in ancient China, once said, "The tree which fills the arms grew from the tiniest sprout; the Residence of nine stories rose from a small heap of earth." With this maxim, Laozi illuminated the concurrence of process and result, as well as the unity of the ideal and reality. By passing through the vicissitudes of the world, a seed grows into a tree reaching up to the heavens. By persistently heaping sand, a pagoda can thusly be built. And by ascending Tai Residence people ultimately have their minds purified, and in doing so understand Dao, or the Way.

茫茫宇宙，大千世界。天地人居中，三才我在内。探天道之奥秘，循地道之格局，结人道之缘分，道法自然，得自在境界，岂不乐哉。

What an immense universe lies ahead! And how kaleidoscopic the world unfolds before us! In between the universe and the world are heaven, earth, and humankind, all of which contribute to the creation of the self. A tripartite self aspires to probe the profundity of the Heavenly Way, abide by the layout of the Earthly Way, and attach to the humanistic spirit. "The law of the *Dao* is its being what it is." Wouldn't it be blissful if the self could reach a higher realm wherein it can be what it is ?

I

形　胜

TOPOGRAPHICAL ADVANTAGES
THE WATCHER OF CIVILIZATION

文　明　守　望

形胜

TOPOGRAPHICAL ADVANTAGES

京畿之地，北通朔漠，南控江淮，右倚太行，左襟渤海，因形胜之佳，建制之妙，历称中都、大都、北平、北京，乃东方文化渊薮，华夏政治中心。

Starting from Beijing, one can move north to the great desert, south to the magnificent Yangtze River, west to the Residence Taihang Mountains, and east to the fathomless Bohai Sea. Therein lies the topographical advantage that Beijing possesses. For this reason, multiple dynasties in ancient China established their capital in present-day Beijing. The great city was known throughout history as the Central Capital, Khanbaliq, Beiping, and so on. Undoubtedly, Beijing is the political center of China and a great aggregator of eastern culture.

文明守望

THE WATCHER OF CIVILIZATION

北京继承了周秦两汉以来的都城旧制，气势恢宏，中轴贯通，左右对称，宫殿居中，前朝后市，左祖右社，坛庙园囿，散置四郊。

九层台恰处坐北朝南、世代护佑的龙脉暨中轴线上，毗邻历代皇都和共和国的心脏，占据京城顶级地标位置，地理形势极其优越。

Beijing's status as imperial capital dates back to the ancient Zhou (1046-256 BCE), Qin (221-207 BCE), and Han (206 BCE-220 CE) dynasties, and it still enjoys an unparalleled magnificence today. A north-south central axis runs through the city, which is renowned for its beautiful bilateral symmetry. Imperial palaces are located in the city's center, in front of which is the central court, and behind which lies the market. To the left of the imperial palaces is the Imperial Ancestral Temple, and to its right is an altar dedicated to gods of the land. In addition, there are numerous shrines, monasteries, gardens, and parks scattered all across the outskirts of Beijing.

Our work – Tai Residence – is located precisely in the great city's central axis, revered as its dragon vein, which has been elaborately protected for generations. The Residence is adjacent to the Forbidden City and the heart of the People's Republic. Situated among the capital city's important landmarks, the Residence's location outshines all its rivals.

极目南天，闾阎扑地，楼影参差；回首北眺，仰山耸秀，奥海腾波。

Looking south along the skyline from the Residence, one sees lanes and alleys crisscrossing beneath the shadows of skyscrapers; to the north, mountains towering aloft and the undulating ripples of the Olympic Lake.

东望鸟巢、水立方各展风姿，奥林匹克建筑群拔地而起；西瞻可见西山如黛，岚气氤氲。

To the east, the Bird's Nest and Water Cube contending in beauty, and the Olympic complex rising awesomely above the horizon; and to the west, one sees the hazy Western Hills, shrouded in upland mist.

II

格 局

SPATIAL LAYOUT
THE INHERITOR OF CULTURE

文 化 传 承

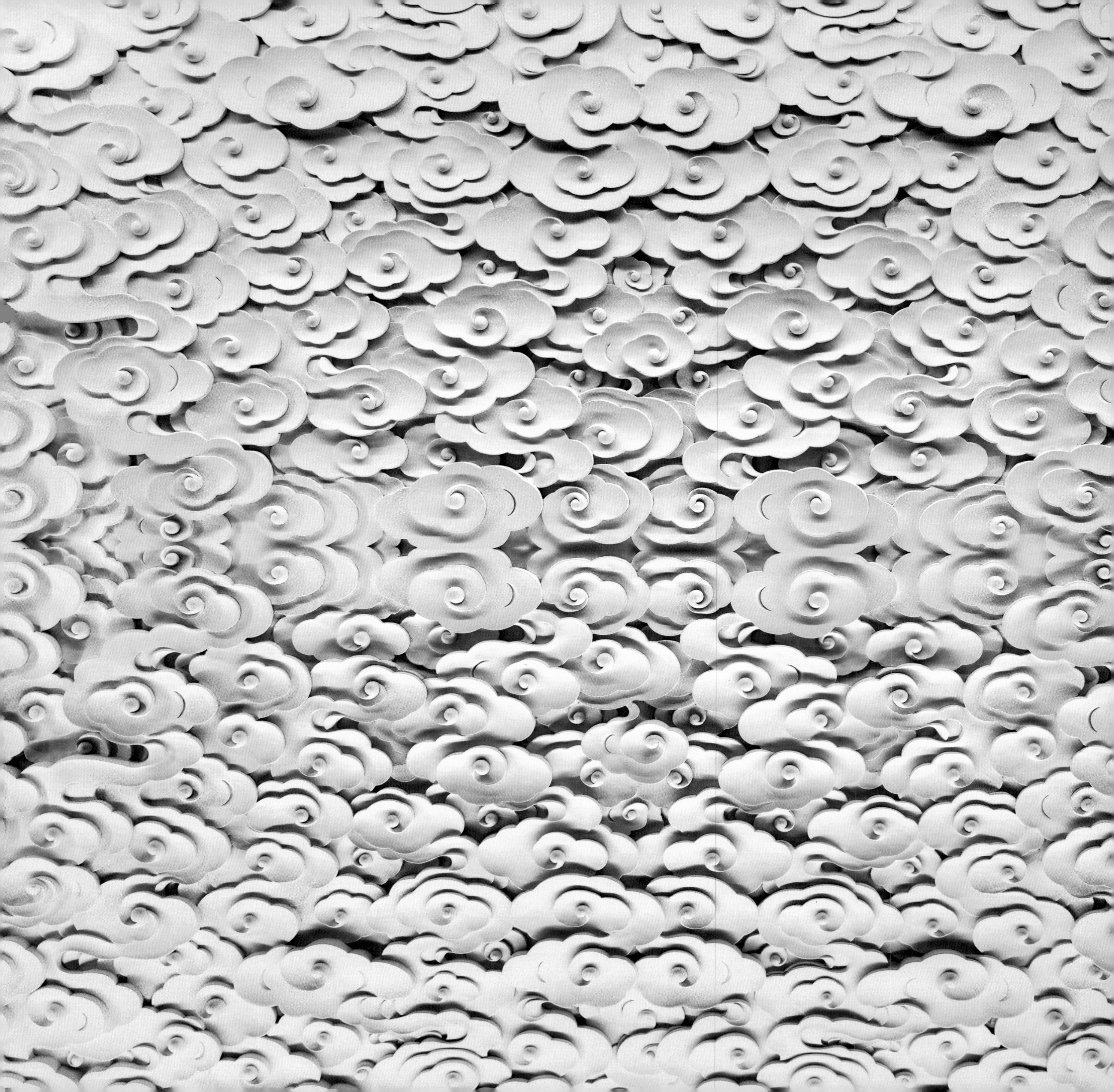

格局

SPATIAL LAYOUT

九层台的主体结构为上、下两层，总面积约为 1,000 平方米。整个建筑结构以交流空间、展示平台、聚会场所、人居环境为基本功能，布局舒展，功能完善，工艺精巧，材质珍稀。移步换形，所见皆为文明守望；腾挪俯仰，触目尽是文化传承。

The superstructure of the Residence is composed of two stories, covering a total area of 1,000 square meters. It aspires to be an architectural complex that satisfies people's needs to communicate, exhibit, mingle, and live. The layout of the Residence is spacious and comfortable, featuring complete functionality. The most advanced craftsmanship was used in building the Residence which is constructed of rare, precious materials. Whenever you go in the Residence traces of Chinese civilization can be found; and wherever you look, you will see traces of Chinese culture.

文化传承

THE INHERITOR OF CULTURE

大门顶天立地，以酸枝木雕刻的蝙蝠铜钱造型，意喻“福在眼前”；扶手采用金镶玉工艺，形成君子事业精进，玉带缠腰之格局。步入前厅，地面采用七彩祥云拼花造型，祥云入境，旺气随轩。墙角配以鸾凤石雕，凤鸣九州；地上铺有云龙地毯，龙腾四海。

The Residence's main entranceway is bold and dauntless. On the door, copper and bat-shaped figures made of rosewood imply that good fortune awaits those who cross the threshold. The golden doorhandle is inlaid with fine jade, signifying progress and prosperity. When stepping into the Antechamber, the colorful, auspicious clouds on the floor greet you. The flying clouds symbolize the success in business. Exquisitely carved, stone phoenixes are positioned at the foot of the walls, signifying that our distinguished guests will amaze the world with their brilliant achievements. The carpet is embroidered with dragons swimming through a sea of clouds, vividly expressing the influence of our dignified guests.

中堂借鉴古代露天听政的布局，背设案几，坐北朝南，接朋会友，既庄严，又开放，平和内敛。背景墙嵌以国家级工艺美术大师王怀俊先生的《四海升平》巨型瓷版画，蛟龙腾雾，不见首尾。中堂两侧用牌坊的设计来取代传统门套，方正挺拔，正气浩然。

The layout of the Presence Chamber is similar to that of the ancient open court for high lords, featuring a fine, south-facing office desk. When granting an interview to visitors, the chamber's grandeur, openness, tranquility, and reserve will impress them. On the wall there is a huge porcelain painting, *Peace Prevailing All under Heaven*, by Wang Huaijun, a nationally esteemed master. Looking up at the wall, a dragon mysteriously weaves in and out of the clouds. The portals at either side of the room innovatively emulate the traditional memorial archway design, resolutely displaying an unswerving uprightness and righteousness.

四海昇平賀盛圖

四海昇平賀盛圖
甲午之春

瑞氣

中堂左侧为荷厅，为主宾交流的饮茶空间。中国文化追求以和为贵，以“荷”喻“和”，体现追求。荷厅以稀有的紫檀大料木雕茶几、沙发布局，所有家具均为工艺大师满工满雕的莲荷图。精雕细刻的荷叶、花蕊与莲枝特写，衬映入室客人品行高雅，谈吐不俗。

To the left of the Presence Chamber is the Lotus Hall, wherein the host and guests of honor can enjoy tea and conversation. As we know, harmony is the emblem of Chinese culture. In Chinese, the pronunciation of "harmony" is the same as that of "lotus". Thus, the Lotus Hall is an embodiment of China's long cherished aspiration for harmony. The tea table is made of precious red sandalwood. Each piece of furniture in the Hall is decorated with the lotus pattern produced by nationally renowned art masters. Silhouetted against the elegantly-carved lotus leaves, stalks and flowers emphasize the refined manners and impeccable morality of the home's distinguished guests.

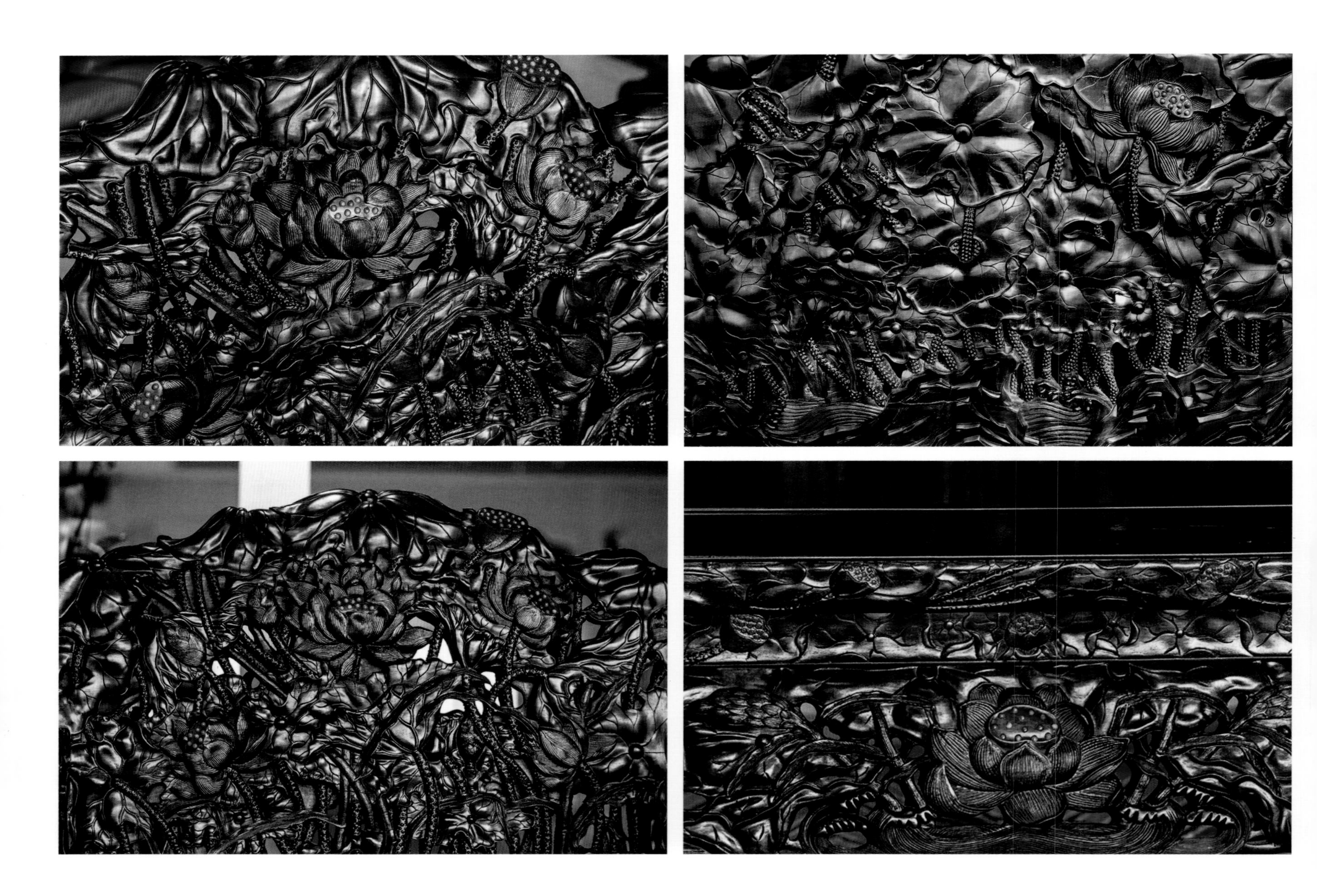

餐厅在中堂右侧。餐厅嵌有“群仙祝寿”“蟠桃盛会”“密宗坛城”的木雕，意为吉祥快乐，借助贵人成就大事。背景以宋徽宗的《瑞鹤图》为主题，采用“四大名绣”之一的“京绣”来表现。仙鹤是高贵的神鸟，寿命千年，长寿富贵。图上彩云缭绕，十八对神态各异的丹顶鹤翱翔盘旋在上空，令人开阔胸怀。

To the right of the Presence Chamber is the Banquet Hall. The Hall is decorated with fine woodcarvings entitled *The Fairy Birthday Celebration, Heavenly Peach Banquet,* and *Esoteric Mandala,* all of which convey that happiness and luck are ubiquitous, and that greatness can be achieved with the help of distinguished guests. On the wall, a reproduced *Auspicious Cranes*, by Emperor Hui (r. 1100-1126) of Northern Song is on display. The painting is a made in the famous Beijing Embroidery style. In traditional Chinese culture, the crane is a sacred bird that can live for more than a thousand years, epitomizing longevity and nobility. Being exposed to the swirling, variegated clouds and eighteen pairs of hovering cranes in the painting, the mind will be substantially broadened.

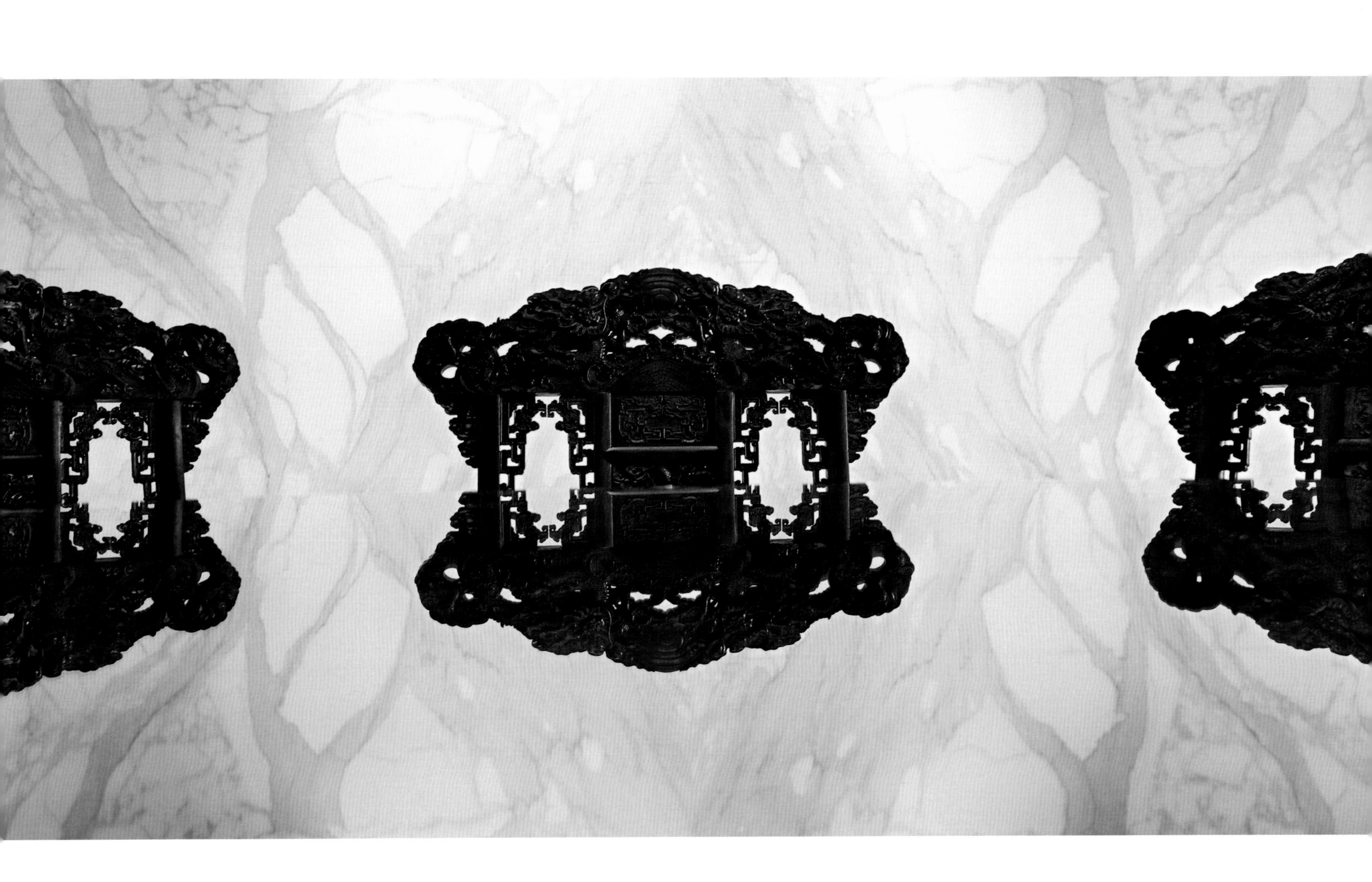

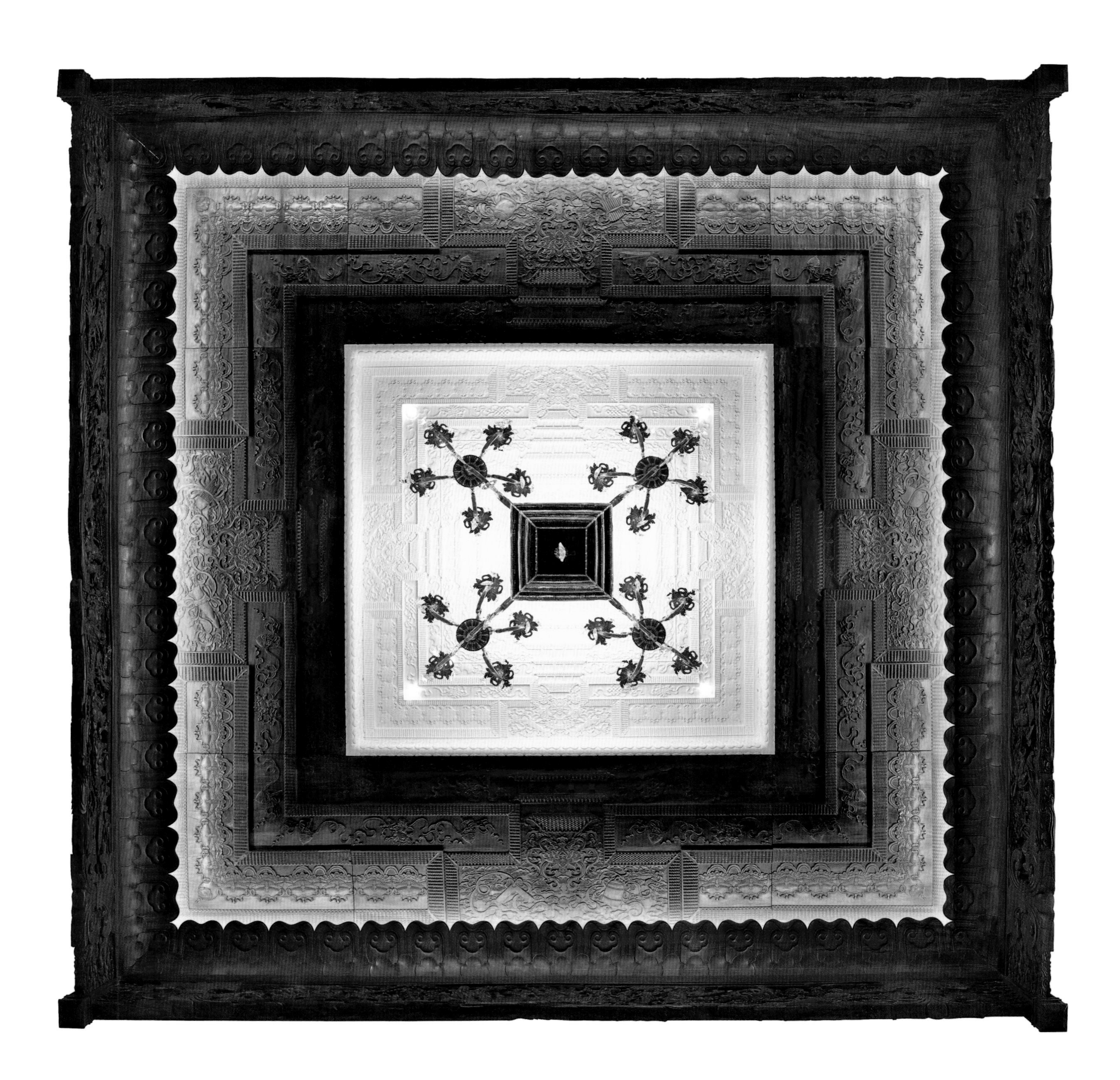

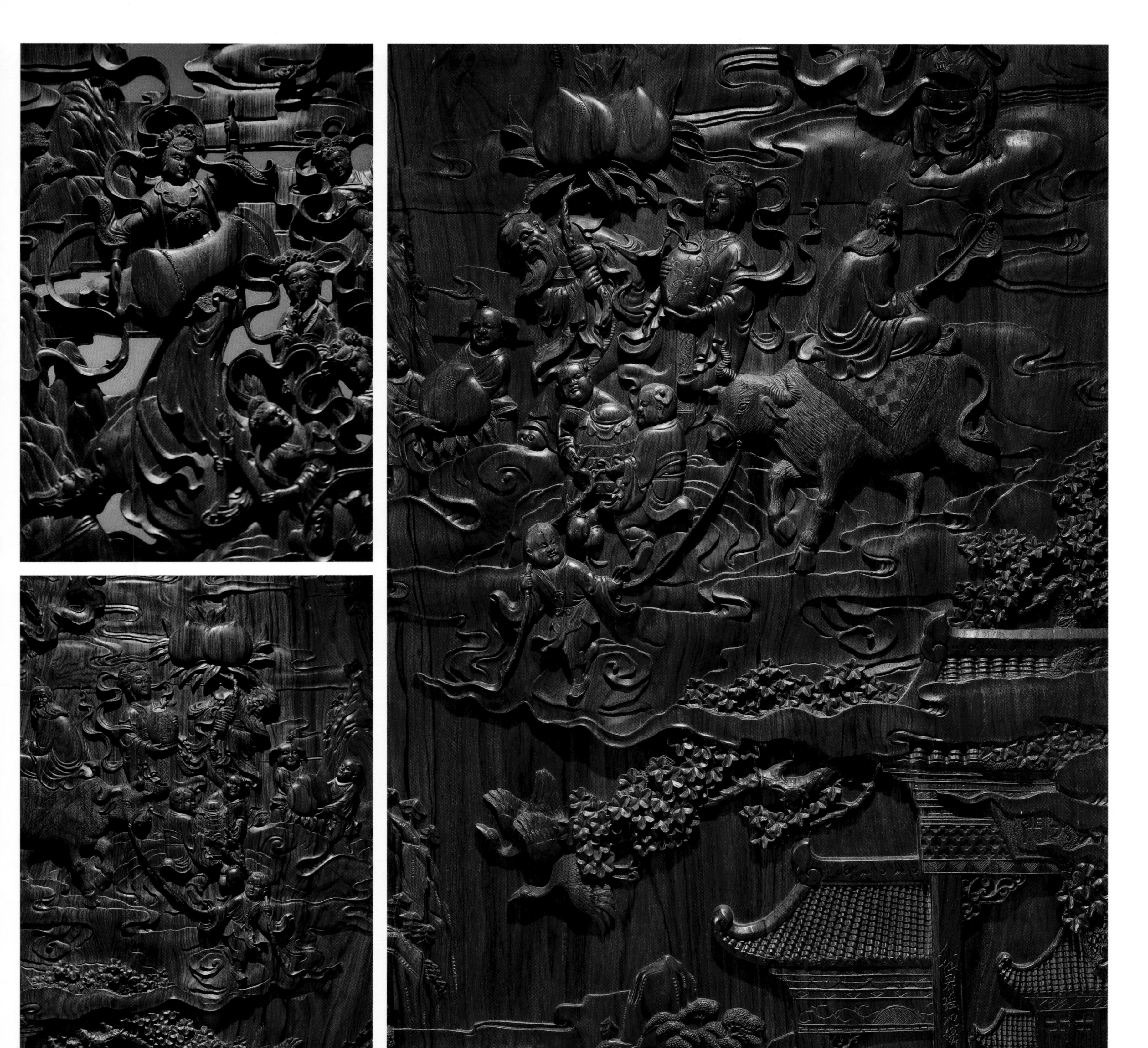

餐厅北侧为休息厅，顶面采用五幅“捧寿”主题的雕刻，有福、禄、寿、喜、财五福齐全之意。博古架上陈设着大大小小几十把造型各异的宜兴紫砂茶壶，客人在品鉴“非遗”传统之余，可随意选择，品茗交流，稍事休息。

The Lounge lies to the north of the Banquet Hall. On the top surface are five birthday-themed engravings, featuring the five blessings – fortune, nobility, longevity, happiness, and wealth. The curio shelves are occupied by dozens of world-renowned Yixing purple clay teapots. Apart from immersing themselves in these elements embodying China's intangible cultural heritage, guests can also sit down and enjoy a cup of top-grade tea.

富贵天下

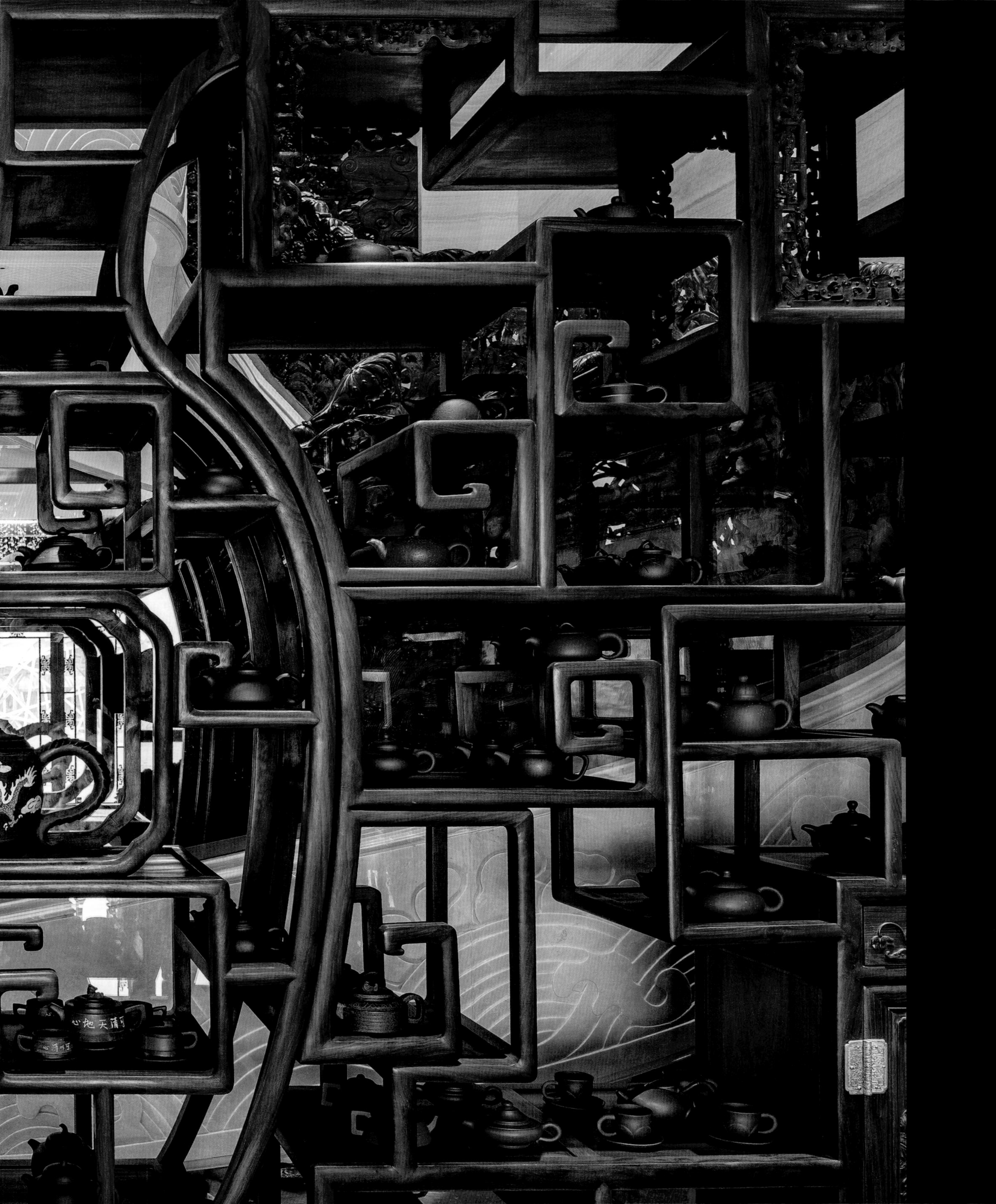

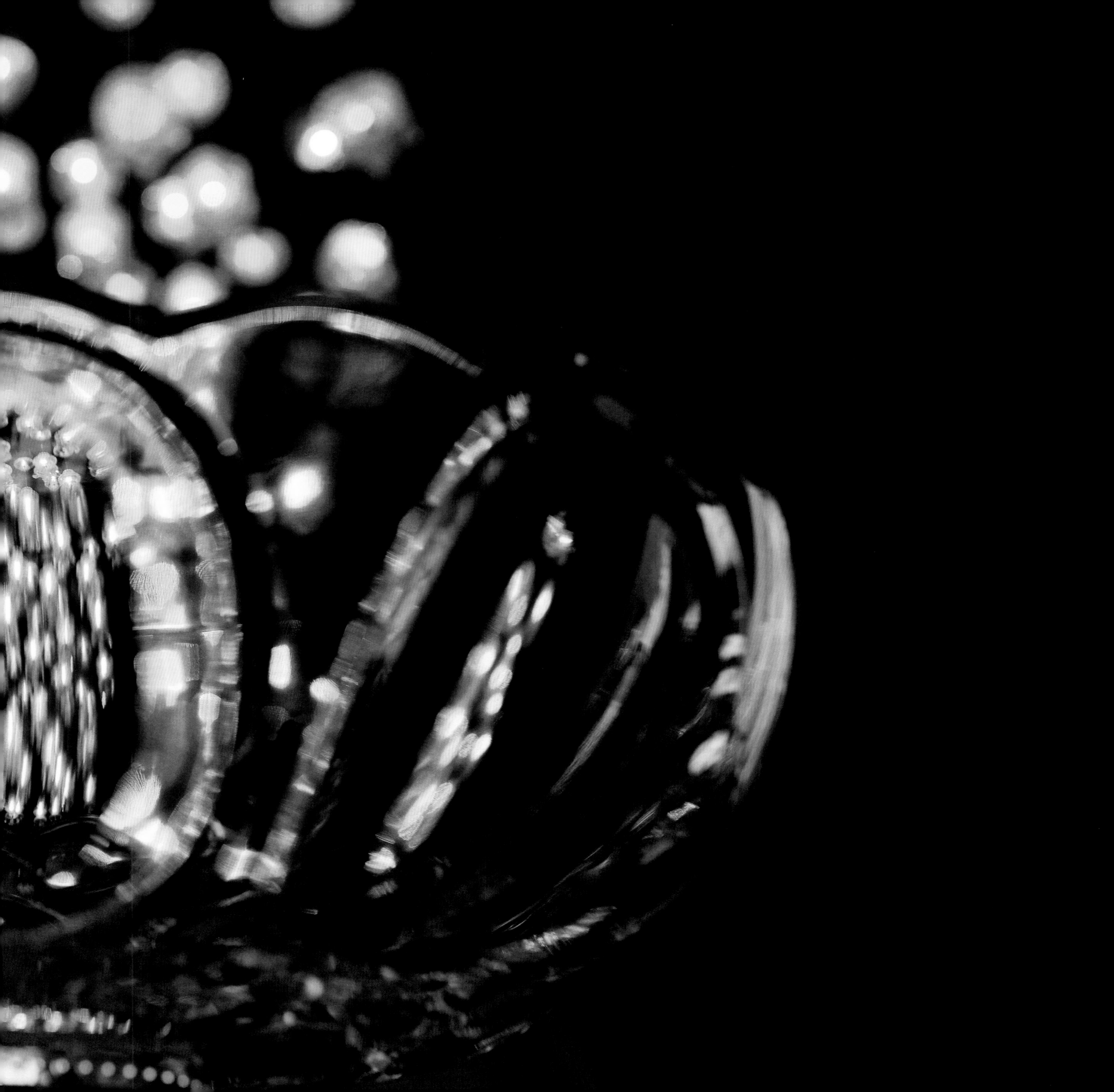

楼层东、西两边设有走廊，东边窗外，鸟巢、水立方尽收眼底；西侧窗外，可极目远眺西山胜景。走廊顶面以柏木雕有“暗八仙”，为八宝吉祥、吉祥平安之意；两侧配以“聚宝瓶”，寓意紫气东来、收纳八方吉祥之气。

Corridors have been constructed at the east and west sides of the floor. Looking through the widow in the eastern corridor, the magnificence of the Bird's Nest and Water Cube unfolds before you; and through the window in the western corridor, the beauty of the Western Hills can be seen in all their splendor. An esoteric woodcarving of Eight Immortals is found on the top surface of the corridors. The recognizable *ashtamangala*, or eight auspicious signs, is a token of perpetual peace. Treasure bowls positioned on both sides of the corridors are actually propitious omens for the accumulation of luck and fortune worldwide.

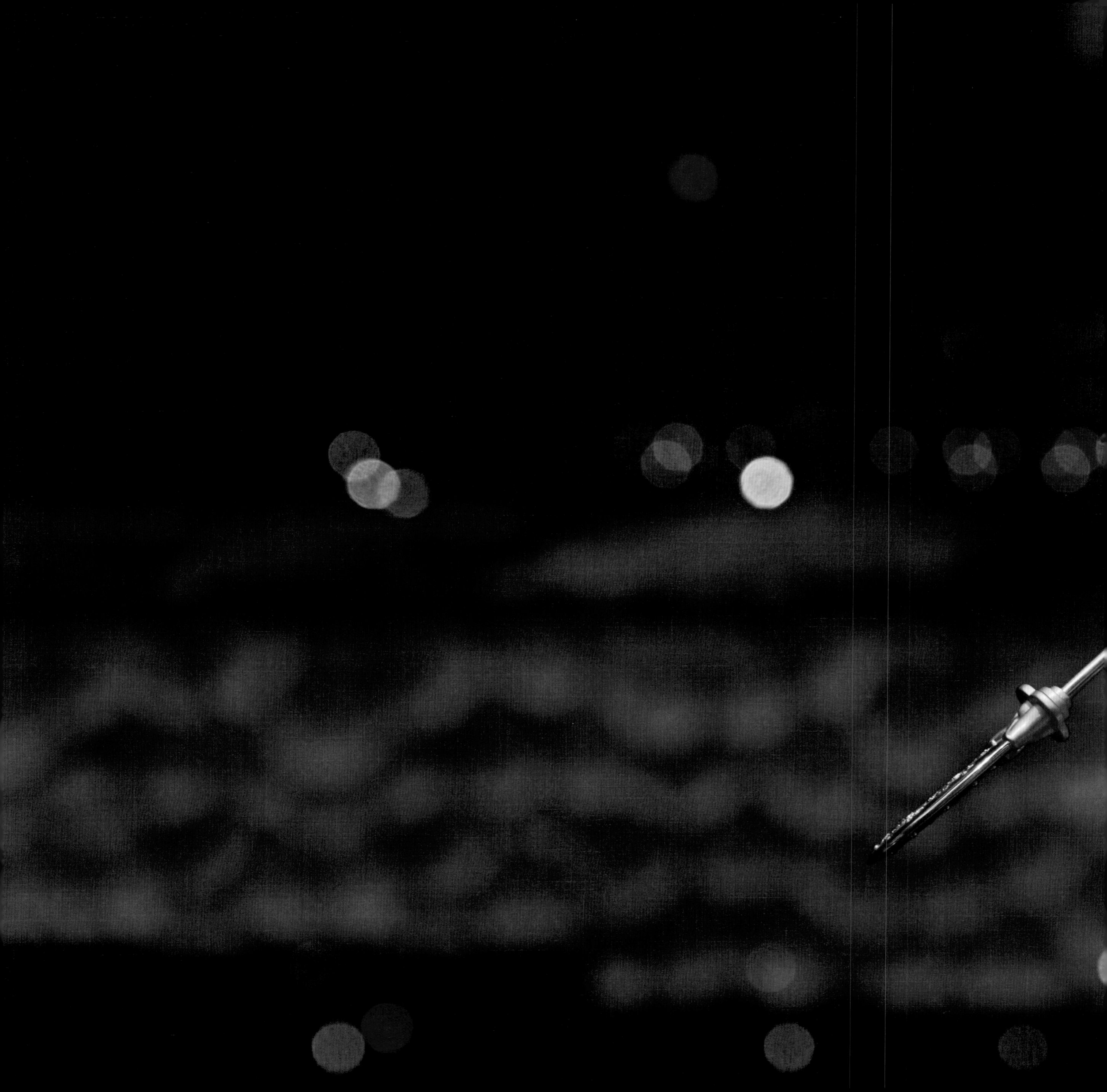

走廊尽头为影音室，又称“畅音厅”，这里巧用囚牛雕刻为主题。“囚牛”是传说中龙的第一个儿子，天生喜爱音乐，用它来雕琢，有传递美妙和谐之意。顶面配以祥云和雨露，呈现出“接天上之甘露，降人间之祥瑞”的景象。以正能量净化整个空间。

At the end of the corridor is a theater room, also known as the Hall of Sonority. The Hall is a brainchild of the designer, inspired by the dragon story. According to a Chinese legend, there was once a Dragon with a musically talented son. A sculpture dedicated to the son represents the harmony that courses through music. The top surface is decorated with patterns of auspicious clouds and dew drops. It is a vivid representation of nationwide peace and prosperity. Overall, the entire Hall is spiritually purified by positive energy.

房顶上的白色水晶灯按青龙、白虎、朱雀、玄武四个方位排列成二十八星宿，用来表现“太极生两仪，两仪生四象”。整体用黑、青、红、金、白五种颜色来代表“金木水火土”五行相生的宇宙观。

The crystal chandelier hung from the ceiling is a display of the twenty-eight constellations positioned in accordance with the Four Directions represented by the four mythological creatures – Azure Dragon, White Tiger, Vermilion Bird, and Black Tortoise – respectively. It epitomizes the sapiential assertion that *Taiji*, or the Grand Terminus, produces the two most basic Forms, and the two Forms produce the Four Symbols (Directions) in *Book of Changes*, one of the greatest books of ancient China. An aggregation of five colors – black, azure, red, gold and white – is applied to the room, symbolizing a profound worldview based on the endless cycles of mutual generation of the Five Elements – water, wood, fire, earth and gold.

一楼为堂，象征人间，散发着温馨缱绻的人性味道和烟火气息。经过池水和石砚为基座，以鱼化龙为主题的木雕楼梯，步上二楼，登堂入室，进入静思妙悟、超越升华的境界。

The ground floor is the hall, symbolizing the human world, wherein lies humanity and earthly happiness. The second floor can be accessed via the fish-morphing-into-dragon-themed staircase, which stands on a pond and a marble base in the shape of an inkstone. The course of ascending to the second floor represents the traditional concept Chinese self-cultivation, whereby humans advance in an orderly way into a realm of tranquil contemplation, sudden enlightenment and ultimate transcendence.

門

登梯后即为“聚贤厅”，豪杰贤能之士可依次落座，畅谈天下，交流古今。聚贤居西侧为议事厅，有完善的会议系统、集展示、交流、研讨、发布于一体，德高望重者同德聚贤，洽谈研判，布局决策。

The Hall of Sages greets visitors on the second floor. In the Hall, those who embody virtue discourse freely about the country and the world, divulging their inspiring understanding of the past and the present. On the left is the Council Chamber, equipped with a complete conference system, whereby exhibition, exchange, discussion, and proclamation can be carried out. What an amazing scenario – in the Hall, persons of lofty virtue and grand prestige gather together and collaborate on ambitious projects.

议事厅配以五层斗拱、景泰蓝吊灯，须弥座托宝瓶。九五宝座两旁放置国家级工艺美术大师赖德全的《源·脉》对瓶，彰显九五之尊的气势。一侧陈列有国家级工艺美术大师张爱廷的青田石雕《十八罗汉》，聚合气场；陈列架顶部至高之处雕有五爪真龙造型，象征成功；蝙蝠木刻满雕造型预示“福满天下”；仙鹤置于厅堂两边，活灵活现。

The Council Chamber is decorated with five-layer interlocking wooden brackets, cloisonné chandeliers, and precious vases on Xumi (Sumeru) stands. Twin vases – *Origin* and *Pulse* – created by Lai Dequan, a master of fine porcelain, are placed at both sides of the Seat of Honor, manifesting the unmatchable prominence of the Guest of Honor. A magnificently carved piece of stonework – *The Eighteen Arhats*, one of the works of the eminent Zhang Aiting – is displayed in the Chamber, generating an air of power. The top of the display stand is engraved with the five-claw dragon, embodiment symbol of success; and the stand, with wooden bats, symbolizes ubiquitous happiness. In addition, the room is flanked by vivid crane statues.

主卧室乾房开敞大气，顶额上“元亨利贞”四个大字明示君子四德。窗外可俯视鸟巢、水立方景观，周边十四条屏木雕葡萄象征硕果累累、兴旺发达；中间镶嵌瓷板画，绘有形态各异的枝头好鸟，呼应木雕的百鸟朝凤图。一树花开，百鸟竞来，一派事业通达、生机勃勃的景象。

The Master Bedroom is spacious and lofty. Four words – *yuan* (Origin), *heng* (Penetration), *li* (Advantage), and *zhen* (Correction) – are inscribed on the door head, illuminating the illustrious virtues of the homeowner, a person of impeccable character. Peering through the window, distinguished guests enjoy the magnificent view of the Bird's Nest and Water Cube. Fourteen pieces of finely engraved, grape-themed wood symbolize success and prosperity. In the center of the room is a porcelain painting depicting birds chirping at the top of tree, working in concert with a carved woodwork portraying a phoenix receiving one hundred birds. Flowering trees and flocking birds denote a flourishing and vibrant career.

元亨利貞

有

元亨利贞

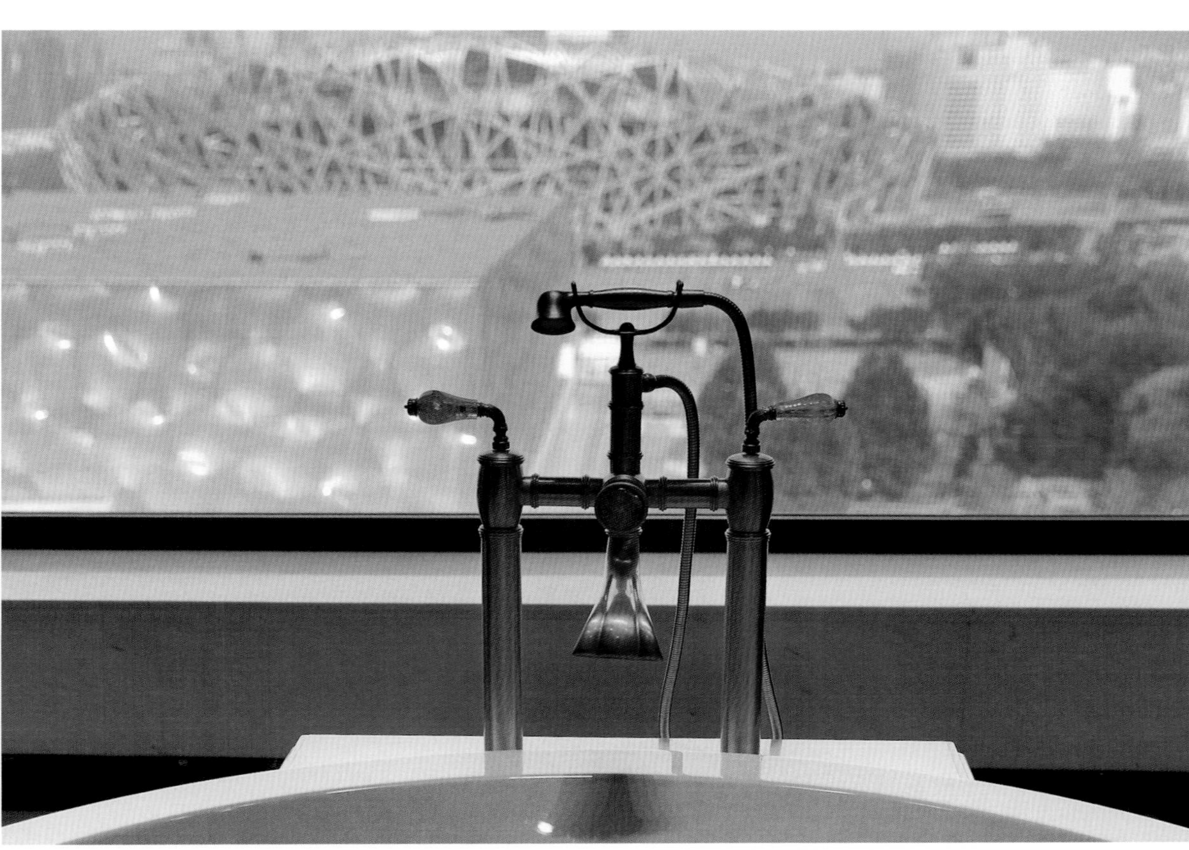

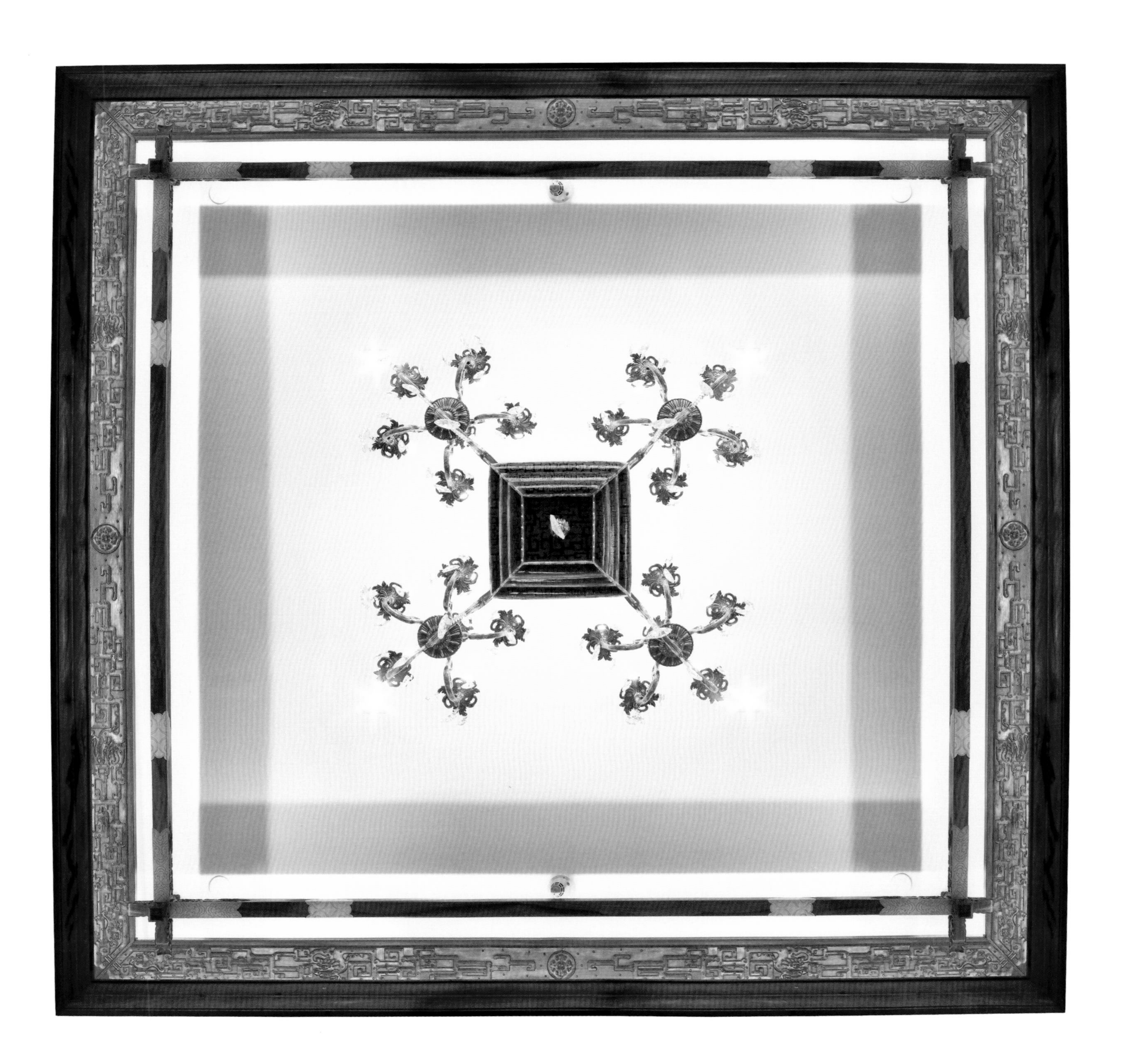

坤房为女卧，通往坤房的乾坤门，一边雕有鲲鹏，一边雕有狮子。进入坤房，可见“黄裳元吉”四个大字，与乾房呼应，寓意做人要用心、坦诚、忠诚、谨言慎行，这才是和谐的最高境界。

The Secondary Bedroom is behind the Gate of Heaven and Earth, which is flanked by kun (a legendary roc) and a lion. Echoing the Master Bedroom, another four words – *huang* (Yellow), *chang* (Lower garment), *yuan* (Origin), and *ji* (Auspice) – can be found here. These characters imply that in order to reach the highest level of harmony, one must exhibit diligence, candidness, loyalty, and cautiousness in conducting oneself.

正

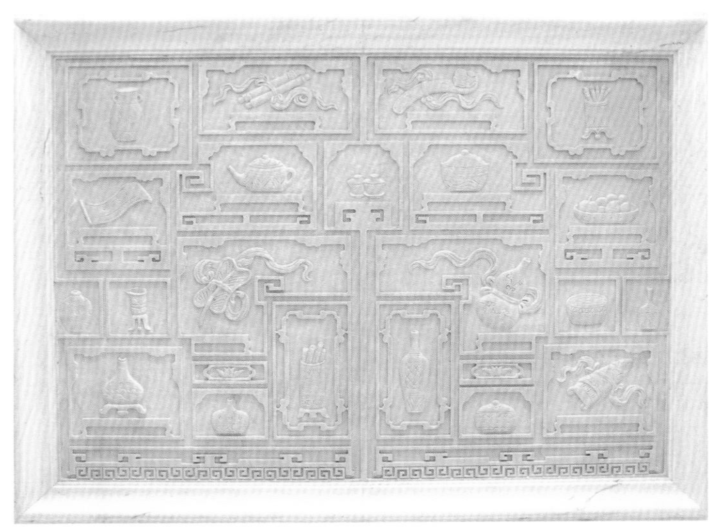

书房为主人运筹帷幄，读书思考之地。书架上左图右史，书桌旁琴剑画缸，几支梅花，一缕沉香，体现了主人的气质和情操。

In the Study, the homeowner may indulge in reading and deep rumination. Halving the shelves, history books and illustrated works greet the prestigious reader. A sword, zither and cylinder (for painting) are anchored to the desk. There are also branches of blossoming plum, which permeate the air with a sweet fragrance, reflecting the homeowner's impeccable morality and lofty sentiments.

匠 心

UNMATCHABLE INGENUITY
AN EMBODIMENT OF CULTURE

文 心 雕 龙

匠心

UNMATCHABLE INGENUITY

老子乐水，提出“上善若水”，认为崇高的人应像水一样滋润万物，有利于万物之生成，而又不与万物相争，能以超自然的心态俯视世间万物，无为无欲，平静恬淡。

Laozi, a Residenceing figure in the Chinese intellectual world, revered water. His assertion that "the highest excellence is like that of water" implies that a person with lofty ideas must act like water; nourishing the myriad things, helping them grow, coexisting peacefully beside them, regarding them with a transcendental attitude, enjoying full freedom from desire, and embracing the highest tranquility and simplicity.

九层台的设计深得老子三昧，十分重视水元素的运用，欲将三千弱水化为一瓢饮。刚进大厅，即闻潺潺水声。随着脚步深入，水声越来越大，由潺潺而流到汩汩而出，有时如万斛珍珠溅落玉盘，有时似千军万马奔腾而来，令乐山者动容，令乐水者为之雀跃，带来心灵波涛的撞击，引发蓬勃向上的生机。

文有文心，诗有诗眼，建筑设计也有枢纽之处。九层台布局结构中令人叹为观止的亮点首推立于水景之上，连接上、下两层的楼梯枢纽。

Tai Residence gives the perfect expression to Laozi's philosophy, attaching great importance to water. In the Residence, the immensity of water is marvelously incorporated into the small size of a pond. As soon as they enter the lobby, visitors hear the bubbling of running water. As footsteps move, the light bubbling turns into heavy gurgling. From time to time, it sounds as if countless pearls are trinkling onto a jade salver, or as though thousands upon thousands of cavalry are galloping forward. Amidst this constantly changing soundscape, persons of virtue and wisdom are moved so deeply that they grow more spiritually and physically vigorous.

Just as all great works of literature have an apotheosis, so too can one find a locus that embodies the central idea of the entire structure in an architectural work. In light of this, it is safe to say the sparkle of the Residence lies exactly in the central staircase perched above the waterscape.

文心雕龙

AN EMBODIMENT OF CULTURE

中枢楼梯的造型图案由名师设计，演绎中国传统的鱼化龙传说，经数百名雕刻大师耗时一年多，运用中国古代传统木雕、石刻技术，全手工精心雕刻而成，可称为真正的“文心雕龙”。

楼梯是衔接上下空间的通道，在风水中是接气和送气之所在，又寓有步步高升之意。整座楼梯下方是一个硕大的砚台水景，黑色墨玉石材寓意“深渊”，君子进德修业，犹如潜龙藏在深渊中，虽然看似风平浪静，却蓄势待发，生机无限。

Based on a beautiful Chinese legend wherein fish transform into dragons, the Residence's brilliant designer has produced the glamorous pattern applied to the central staircase. Then, hundreds of master craftsmen spent more than one year engraving the pattern into the staircase. The method used by these craftsmen was a very traditional Chinese wood/stonecarving technique. The staircase was finished entirely by hand. It truly is a masterpiece embodying extraordinary design and outstanding craftsmanship.

The central staircase bridges the two floors, becoming a point that inhales and exhales *qi*, or the vital breath, in the geomantic perspective. It is also an embodiment of the traditional aspiration to attain eminence step by step. Right below the staircase is a huge waterscape in the shape of an inkstone. As regards the materials, the black jade implies that the self-cultivation of a superior person is exactly like the dragon's quiet and calm effort to gather strength in a deep gulf.

远看整座楼梯，在“天”“地”乾坤中像一棵千年古树擎天而立，又像一股水浪旋转直上，形成“太极”之势。墙面采用了整块条形白玉，代表土地，也暗喻资源的积累；鱼肚白的纹理做成树根的形式，寓意根深而广袤，源远而流长。当黑色砚台中的水慢慢溢出时，不禁令人联想到“月盈则亏，水满则溢”的规律。

青色的老树根石雕被雕有荷花和鲤鱼的红色酸枝木扶手包裹着，形成千年老树特有的“木包石”，在金色铜底座的衬托下显得更加稳固，象征着成功来源于深厚的根基。

From a distance the staircase looks like a thousand-year-old tree growing between heaven and earth. More philosophically, it is also like a spiral wave as mighty as the Grand Terminus, or the ultimate creator of the myriad things. A whole piece of pure jade is inlaid in the wall, symbolizing the earth and the accumulation of resources. The calacatta texture is turned into a pattern emulating the root of a tree, signifying the strong attachment to tradition and origin. As water slowly rises in the huge inkstone made of dark jade and eventually spills over, it remind us of the insurmountable natural law that everything has its limit.

A rosewood piece engraved with lotus and carp encircles a dark blue tree root stonework, creating a masterpiece of *Stone Enveloped in Wood*. Standing on a gilded bronze base gives it a look of strength and solidity. The work symbolizes that a solid foundation is a requisite for success.

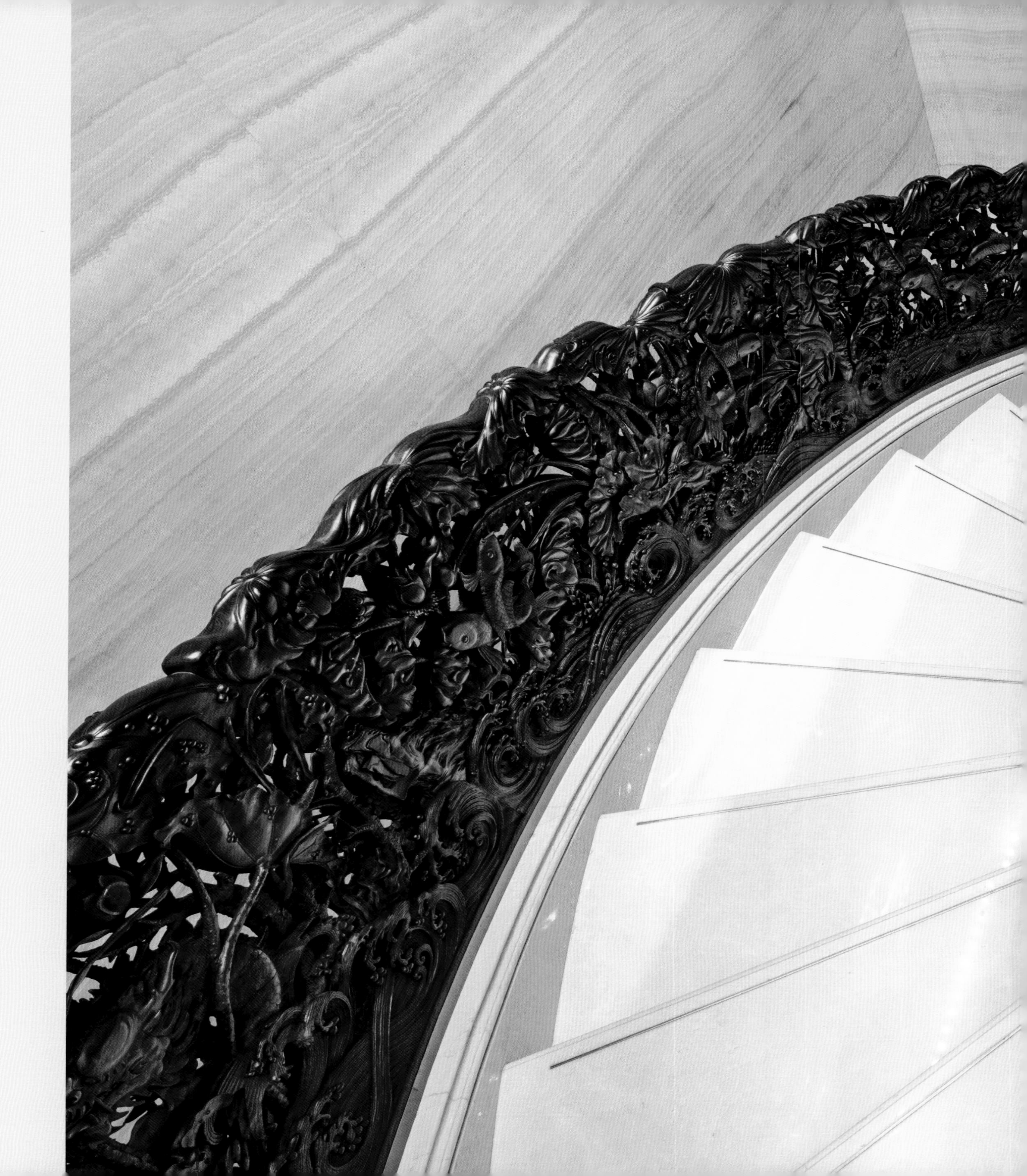

墨玉石材“深渊”之上是水墨玉石雕刻的翻腾水浪，交错着小叶紫檀、红酸枝雕刻的荷花、螃蟹、螺蛳、水草和形态各异的鲤鱼，呈现出鱼戏荷叶间的融洽氛围。“荷”与“和”谐音，寓意“和气、平和”；“蟹”与“谐”谐音，寓意“和谐、欢谐”。水，万物之源；石，开天辟地；木，生生不息，三者之间精妙搭配，携带着孕育万物的原始自然力量。

A dark-jade stonework depicting surging tides is placed above the inkstone, intermingling with lotus, crabs, snail, water plants and carps of every hue, which are made of rosewood and sandalwood. They all form to create an impressive scenario in which fish are swimming happily among lotus leaves. In Chinese, *he* (lotus) refers to harmony; and *xie* (crab) denotes concord. When *he* and *xie* combine, there is *he-xie* (a harmonious world). Moreover, water stands for the origin of the myriad things; stone, the tool whereby the world is built; and wood, the endless propogation of all sentient beings. The marvelous trinity of water, stone and wood sheds revealing light on the existence of the most primordial and creative force, which gives birth to all.

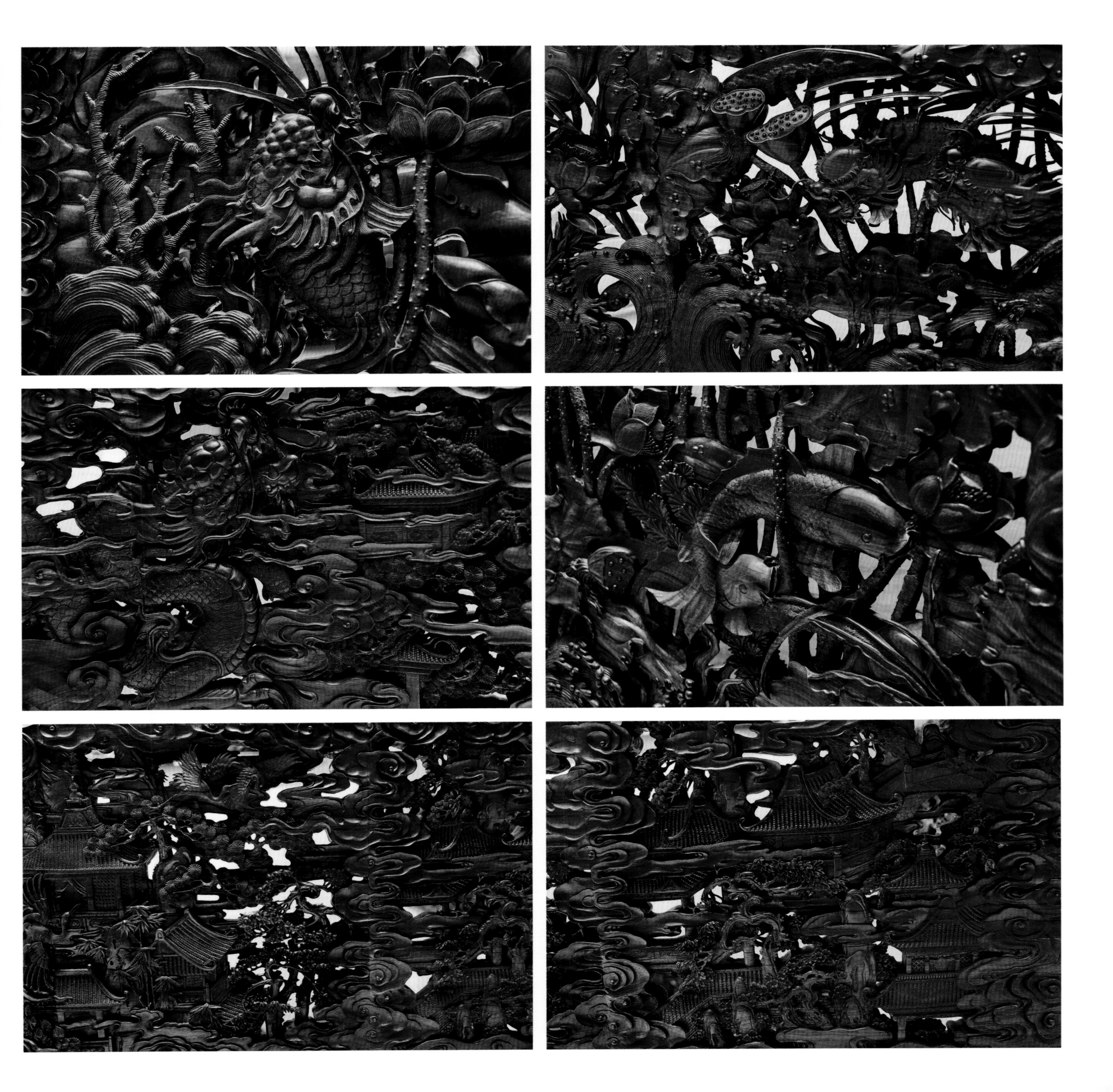

楼梯整体造型演绎的是中国古代传说黄河鲤鱼“耐守寒潭未济中，不觉一朝头角耸，禹门一跳到龙宫”的故事，传递了“鱼为奔波始成龙”的变化思想，融汇了“龙图腾”、鲤鱼跃禹门图，松柏、仙鹤等多种祥瑞事物于一体。青云直上，幻化为龙，云龙交融，显示出一种似云非云，似龙非龙的神秘图案，体现出中国文化的意象之美。

数百名雕刻大师凭借丰富的雕刻意境、娴熟的雕刻技法，通过圆雕、浮雕、透雕、嵌雕、通雕等雕刻技术，以细腻的刻画、流畅的线条，把龙的威武、鱼的活泼、水的灵秀、云的飘渺、风的清爽、荷的柔美表现得栩栩如生。

The central staircase is a perfect architectural narration of the legend that carps in the Yellow River had to swim arduously before finally leaping into the Dragon Palace. The legend metaphorically places emphasis on the importance of *change*. Artistically, the staircase is a combination of the dragon totem, leaping carps and auspicious omens such as pines, cypresses and cranes. As regards the visual effect, the interlaced patterns of dragon and clouds create mysterious figures. The work well exemplifies the beauty of imagination found in Chinese culture.

It is particularly worth mentioning that hundreds of master craftsmen spent enormous amounts of energy and time creating the abovementioned wood and stone works. Depending on their unparalleled perception of sculpture, skills, and techniques, the craftsmen vividly, exquisitely, and smoothly depicted the might of the dragon, the vitality of fish, the delicacy of water, the ethereality of cloud, the briskness of wind, and the morbidezza of lotus; all by means of a kaleidoscopic array of traditional and modern engraving processes.

整座楼梯上方的水晶灯是按照密宗二十八星宿排列而成的，星星点点，包罗万象，只可仰望而不可亵玩。一盏盏明灯如同宇宙星云里的一颗颗恒星，华灯齐放，熠熠生辉，亮如白昼；灯光熄灭后则如夜空苍穹，寂静辽阔。二十八星宿水晶灯笼罩万物，俯视众生，与下方的水墨砚台交相辉映，构成一天一地、一乾一坤、一阴一阳、一实一虚的意境。

精湛的工艺，丰富的造型，无不体现缜密、繁复、细致、典雅的风格，精雕细琢、鬼斧神工般的杰作让人目不转睛，触动心灵、浮想联翩，似沉醉于蓬莱仙境一般。

The crystal chandelier on top of the staircase is positioned in accordance with the esoteric layout of the twenty-eight constellations. Its twinkling crystals imitate a starry night sky, embodying the myriad things and enjoying total freedom from even the slightest vulgarity. In this image, the bright lamps of the chandelier make up the stars of the sky. When lit, the room is as bright as day; when the lights are switched off, a vast, silent night sky unfolds before you. The twenty-eight-constellation-themed chandelier overlooking all sentient beings and the dark-jade inkstone lying below complement each other's beauty. Taking these features in, visitors perceive heaven and earth, men and women, and existence and nonexistence in the subtlest and most profound way possible.

The exquisite craftwork and abundant sculptures are all perfect embodiments of the Residence's deliberateness, complexity, meticulousness and elegance. All such works are masterpieces produced by extraordinarily talented maestros. Gazing at them, everybody – without exception – is moved, and their boundless imaginations are unleashed, as if entering an edenic paradise.

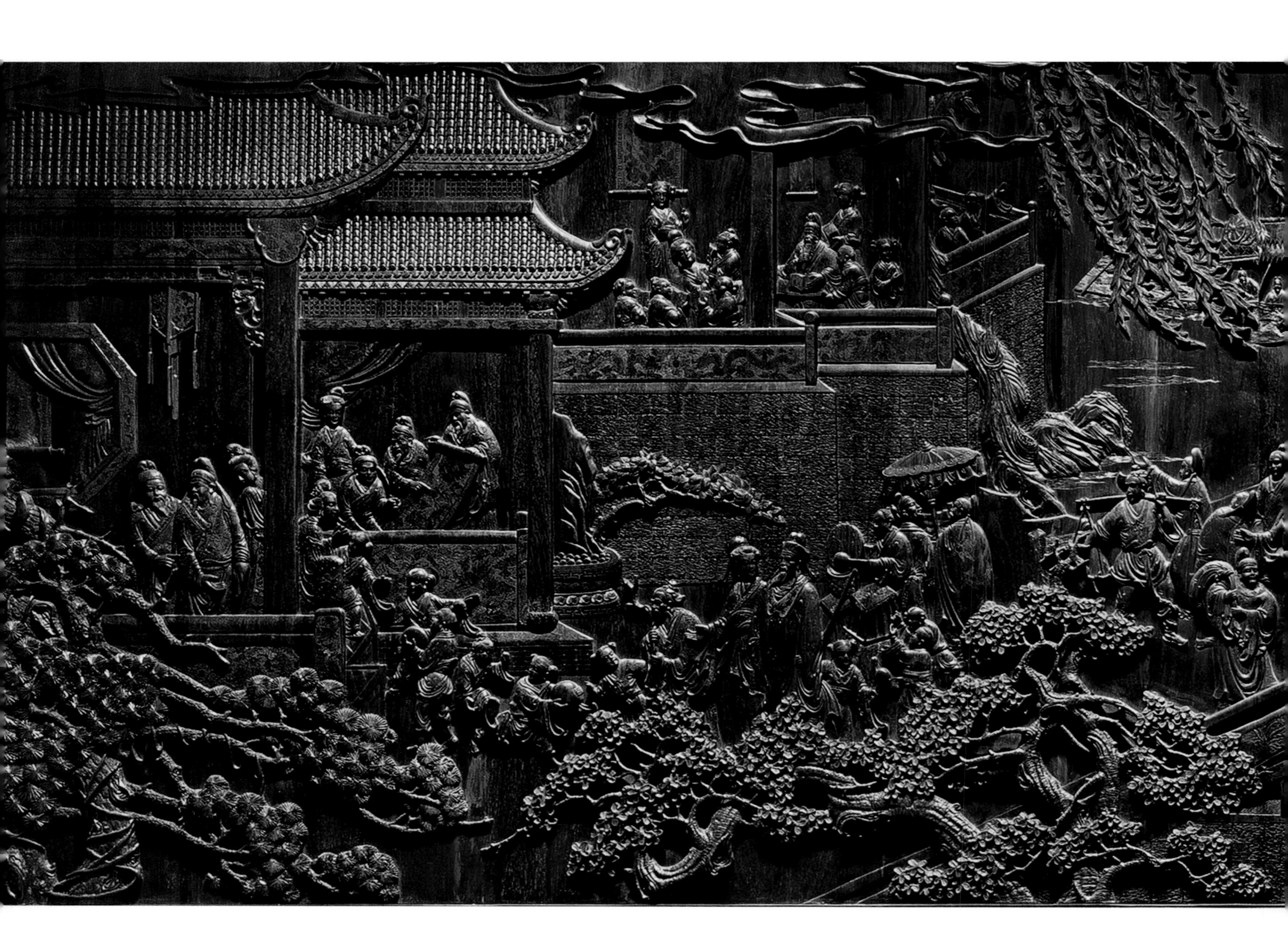

九层台的装饰处处精品，事事匠心。瓷板画、水晶灯，精品卫浴，顶级音响，无一不绝。其中紫檀雕刻技艺特别突出，蟠龙腾云，醒狮巡山，人物故事，花鸟图案，在每个房间都触目可见。鸟类的雕刻手法是面临失传的丝翎檀雕，把鸟的神态和羽毛、花卉的枝叶和苞蕾雕刻得精细、逼真。这些紫檀雕刻精品栩栩如生地刻画出了植物和动物的生命精神，体现了中国雕刻艺术的本质属性。

The Residence's decorations are all finished in accordance with the most demanding standards. Porcelain paintings, crystal chandeliers, sanitary equipment, and modern acoustics are all uniformly top-notch. The rosewood woodcarvings are particularly outstanding. They can be found in every room, covering highly diverse themes, such as dragons flying in clouds, awakening lions among mountains, historical figures, and flowers and birds. When depicting birds, art masters employed a rare technique uniting the traditional Chinese painting and basso-relievo. Thanks to such a novel approach, the delicate mien and in particular the fine feathers of the birds and branches, leaves and buds of flowers are portrayed with exquisite verisimilitude. The Residence's rosewood works are indisputable masterpieces articulating the vitality of plants and animals. They are no other than perfect incarnations of the Chinese art of sculpture.

IV

理　念

GRAND IDEAL

A TRANSMITTER OF DAO

文　以　载　道

理

GRAND IDEAL

九层台是以高端人士的交流、展示、休闲、居住为主要功能的活动空间，其设计理念来自被誉为“群经之首”“大道之源”的《易经》和百家典籍。整个布局结构和景观装饰将中国传统文化中的“天地人”三维格局、“仁义礼智”四德品行、“金木水火土”五行观念，以及阴阳乾坤、天人合一的思想融为一体。

The Residence is a place where the social elite live, rest, display themselves, and associate with each other. The most basic idea guiding the founding of the Residence lies in the *Book of Changes*, which is revered as the *classic of classics*, and the source of Chinese wisdom. Of course, works produced by masters of various intellectual schools in ancient China also contribute to the guiding ideology. Overall, structurally and visually, the Residence seamlessly combines the most fundamental elements of traditional Chinese culture – the trichotomization of everything into heaven, earth and humankind, the Four Virtues (benevolence, righteousness, propriety and wisdom), the Five Elements (metal, wood, water, fire and earth), the duality of *yin* (negative/dark) and *yang* (positive/bright), the unity of heaven and humankind, and so on – into one.

九层台的件件墙饰木雕如温和敦厚的长者，缓缓讲述着华夏历史的故事和智慧。在吸纳传统经典精华的基础上，又加入了人生的历练与感悟。《孔子学艺》《观云思亲》《孔融让梨》《苏武牧羊》《诸葛进表》，几乎每一处景观都有史可稽，有据可考。

九层台的设计深得五行有机论思想的玄妙。金木水火土，既循环相生，又循环相克，圆满之势，欣欣而生，由此构成一个有机整体，建立起“金木水火土”五行相生相克的宇宙观和生命哲学，其间蕴涵的有机论思想成为中国古代科学认识论的精华。

Like a gentle, venerable elder, the woodcarvings embellishing the walls of the Residence slowly and steadily narrate fascinating stories about Chinese wisdom and historical enlightenment to all visitors. Such stories embody classics, history and life. All of them – *The Knowledge-Seeking Confucius, Longing at the Sight of Floating Clouds, Kong Rong Giving up Pears, Su Wu Tending Sheep, Zhuge Liang Presenting a Memorial on the Eve of the Northern Expedition*, for instance – are grounded in historical fact.

The mutual generation and overcoming of the Five Elements was one of the intellectual fountainheads of the Residence's lead designer. In endless cycles, the five most basic materials which make up the world – metal, wood, water, fire and earth – simultaneously promote and restrict each other. It is in such transmutations that perfection is achieved and an *organic whole* created. Ancient Chinese cosmology and life philosophy were based on the idea of the Five Elements. The *organism*, one of the key constituents of traditional Chinese cosmology and life philosophy, is one of the highly inspiring epistemological assets that ancient China contributed to the world.

文以载道
A TRANSMITTER OF DAO

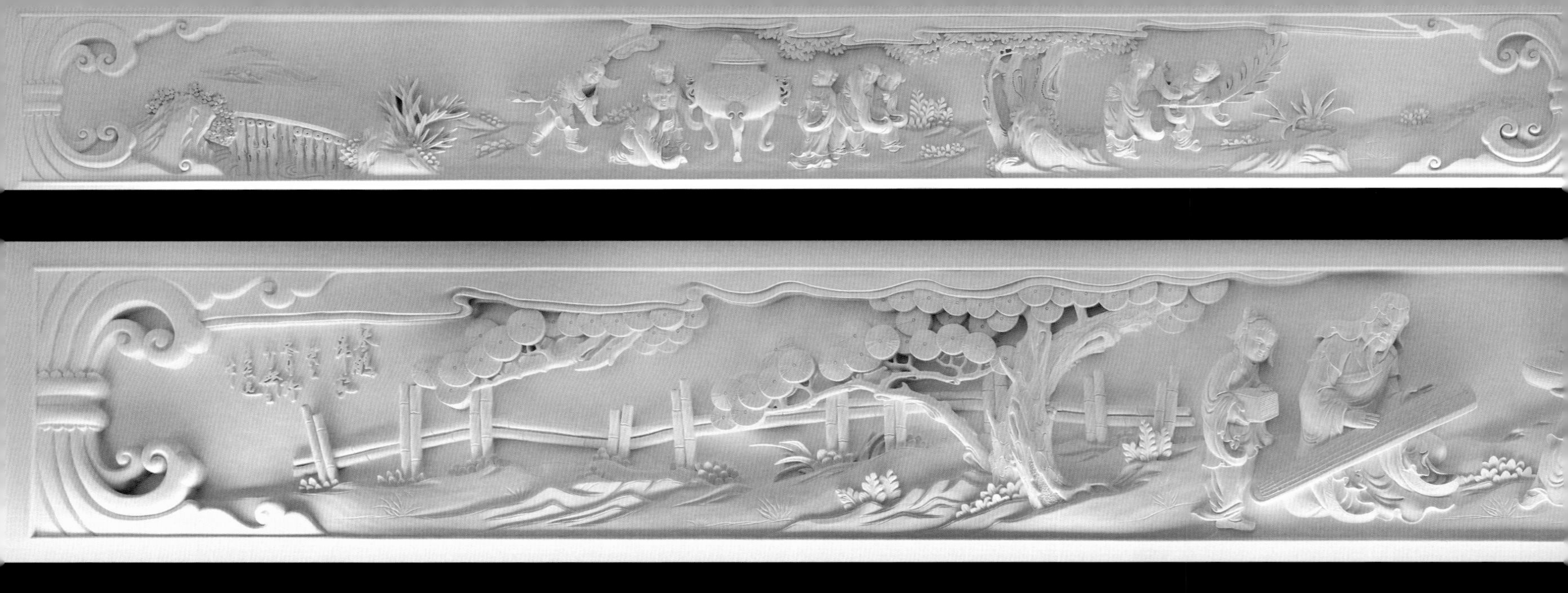

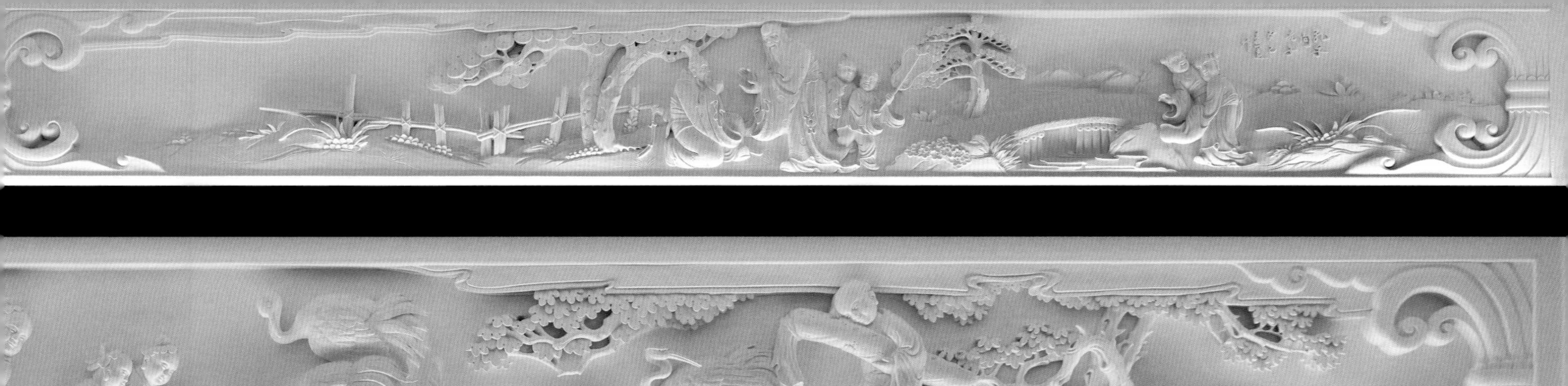

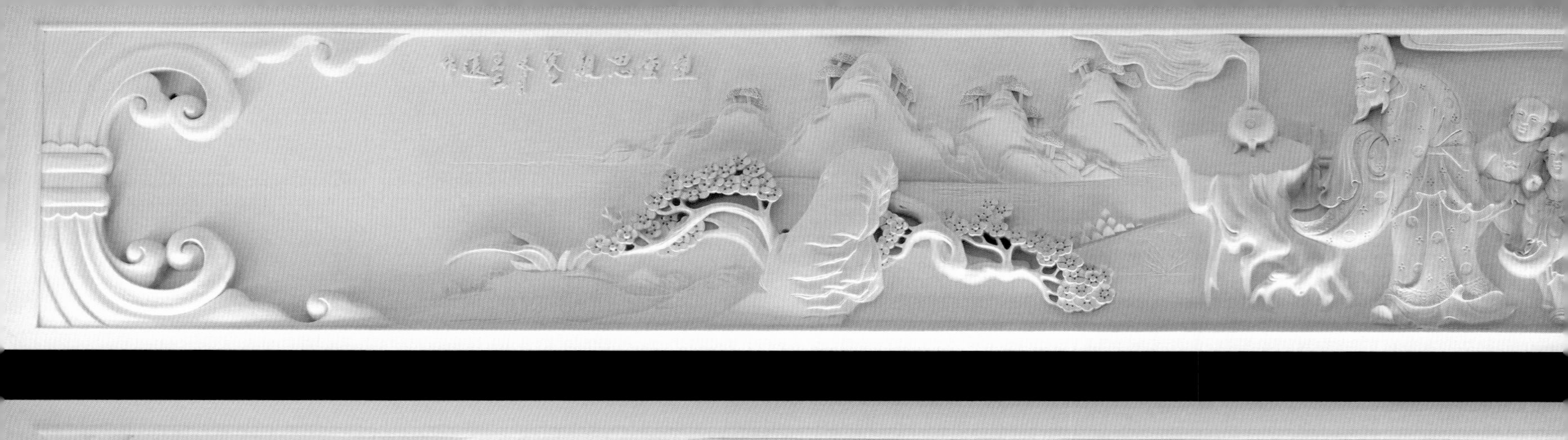

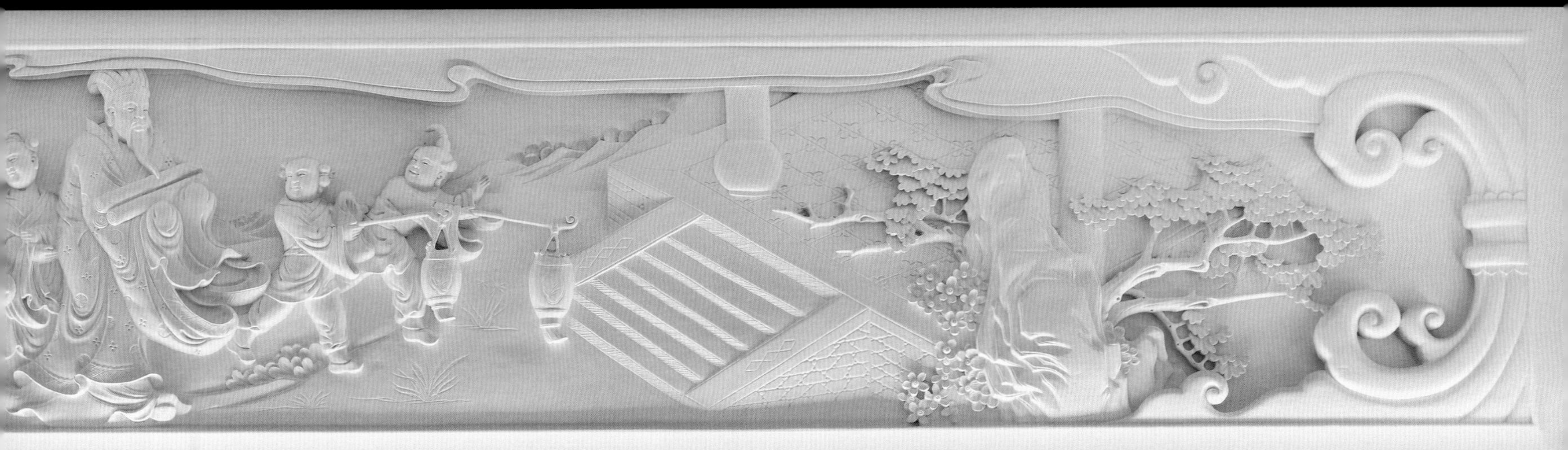

“九层台”之名源于老子。在老子心中，水有七善，“居善地，心善渊，与善仁，言善信，政善治，事善能”。九层台上、下两层的枢纽立于水池之中。水虽然柔弱，却可以洗去身体的污秽，物品的污点，具有变幻莫测的美感，澄澈透亮的风韵。人要学水适应环境，以超自然的心态俯视世间万物，无为无欲，平静淡然，自然进入从“无为”到“无不为”的境界。

Tai Residence's appellation comes from *Laozi.* According to Laozi, water is a multipurpose benefactor; it benefits the home, the mind, nature, words, governance, and ability. For this reason, it is in the pond that the central pivot point of the Residence was erected. Though water is gentle, it can wash the filth from people and objects. Water is clean and clear, and fascinatingly unpredictable. Water has long inspired humankind to be adaptable to changing circumstances, and it has taught us to hold ourselves aloof from excessive material desires, instead urging us to enter the realm wherein everything can be done through nonaction.

水生木。人类自古以木为伴。红木是大自然赐予的宝贵财富，不仅纹理细腻、坚硬耐磨、手感润滑，非常适合雕刻，而且具有高品质的稳定性，随着时间的沉淀，会更加饱满细腻，即使几百年后，只要稍加润泽，依旧焕然若新。九层台使用紫檀、酸枝雕刻的《蟠桃祝寿》《福满乾坤》《凤鸣九天》《龙腾四海》，不仅呈现木材本身的生命之美，而且彰显出中华传统文化的博大精深。

Water generates wood. Humankind has enjoyed the close companionship of wood since ancient times. Rosewood, one of the most precious gifts bestowed by Nature, is renowned for its abrasion resistance, smooth texture, stability, and durability. Most importantly, it is very suitable for sculpture. Rosewood also stands the test of time, for as it ages, it will acquire a beautiful, glossy patina. Even after hundreds of years, it looks new, as if it were lightly moisturized. For exactly this reason, the Residence exclusively applies sandalwood and rosewood to its woodcarvings, such as *Heavenly Peach Banquet, Jubilant Universe, Phoenix Singing to the Nine Heavens,* and *Dragon Flying over Four Seas.* These masterworks showcase the vital beauty of wood and represent the broadness and profundity of traditional Chinese culture.

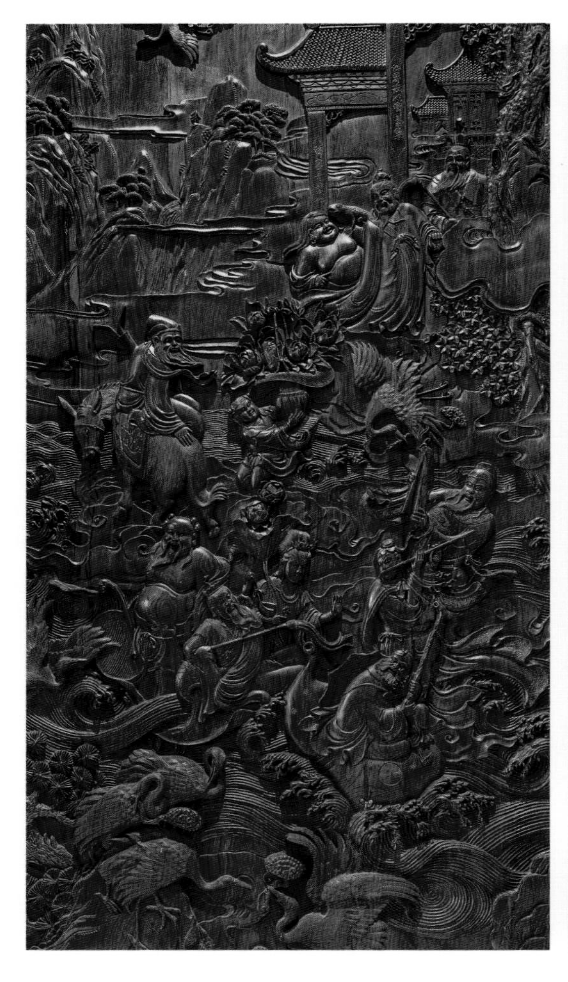

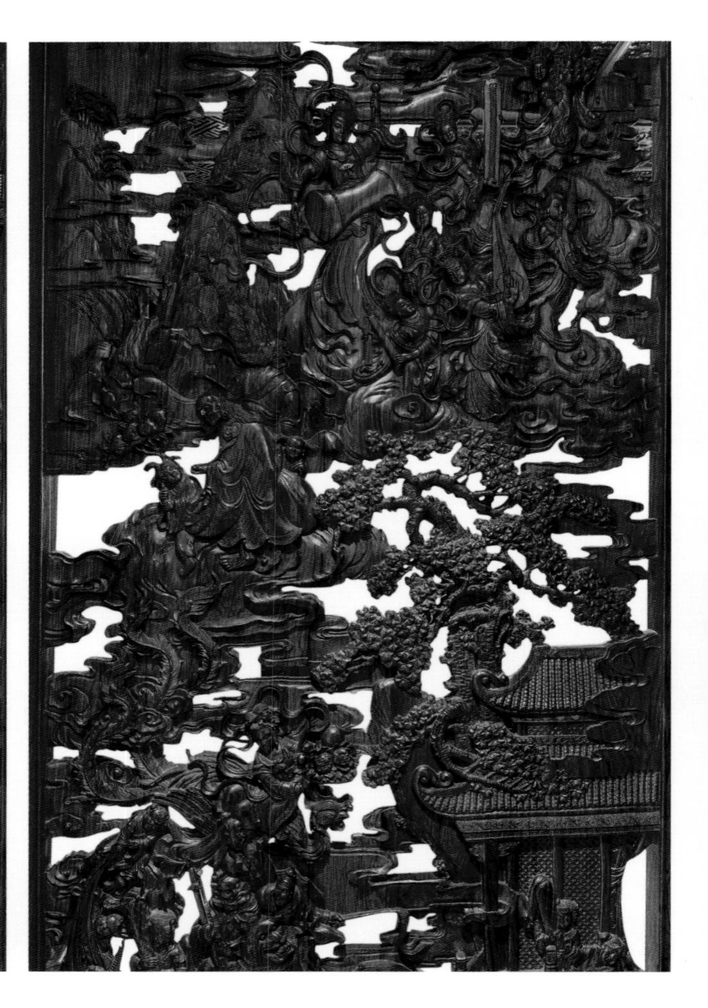

木生火。经过高温窑火烧炼的景德镇瓷板画属火。作为国家级非物质文化遗产代表作项目的瓷板画，在明中叶开始出现，嵌瓷屏风无论是围屏、插屏还是挂屏，都经常会见到上边镶嵌有装饰意味浓厚的瓷板画。九层台的瓷板画均出于当代陶瓷国家级工艺美术大师之手，王怀俊的《四海升平》，徐子印的《百狮图》，以及装饰于不同卧室空间的十几幅《花鸟图》，寓意深刻，釉色鲜艳，书画相衬，工艺上乘。

Wood generates fire.The porcelain paintings produced in Jingdezhen's high-temperature kilns are of a *fire* nature. Porcelain painting emerged in China as early as the mid-Ming dynasty (1368-1644), and it has become an invaluable national intangible cultural heritage. The highly decorative porcelain painting is widely applied to folding screens inlaid with porcelain. The paintings of the Residence were all created by nationally-esteemed maestros. *Peace Prevailing All under Heaven*, for instance, was produced by Wang Huaijun; and *Hundred Lions*, by Xu Ziyin. In addition to the brilliant works, a dozen porcelain paintings depicting birds and flowers can be found in the Residence. These paintings, all of which embody fine calligraphy and traditional Chinese painting, are profoundly meaningful, brightly glazed, and extraordinarily superior.

獅坐江山

宏贯天下

火生土。一身多彩，由火山岩转化而来的浙江青田石属土。九层台收藏有国家级工艺美术大师张爱廷的青田石群雕《十八罗汉》。大师以十八罗汉为主题，因材施艺，依色取巧，将神态各异的罗汉形象与石材原有的异色斑痕巧妙结合，使石材的缺点转化为艺术作品上的点睛之笔，同时令所塑造的人物形体也具有不同的色彩变化，堪称巧夺天工的国家级非物质文化遗产。

Fire generates earth. The colorful Qingtian stone from Zhejiang, one of the primary materials of stonework used in the Residence, is of an earth nature on the grounds that it is a type of volcanic rock. One of the Tai Residence's landmark stonework pieces, entitled *The Eighteen Arhats*, is made of Qingtian stone. This piece is one of the works of Zhang Aiting, a nationally-recognized master of art. After researching the original shape and color of Qingtian stone, Zhang marvelously carved the Arhats with different bearings, and the original black spots of the stone have been perfectly fused together. Moreover, he has endowed the Arhats with subtle changes in color. By doing so, the shortcomings of Qingtian stone is artistically – almost magically – turned into something laudable. Zhang's masterpiece is no other than a result of divine skill.

土生金。作为国家级非物质文化遗产代表作项目的金属工艺品景泰蓝属金。景泰蓝的制作既运用了青铜和瓷器工艺，又融入了传统手工绘画和雕刻技艺，堪称中国传统工艺的集大成者。景泰蓝曾经是皇宫大殿的主要陈设，紫禁城金銮殿、国子监辟雍宫、颐和园排云殿，在这些皇家专属的殿堂里，景泰蓝的宝石般光芒总是令人赞叹不已。九层台上、下两层均布置有景泰蓝的器物陈设和装饰，八宝尊的华贵，大座钟的沉稳，灯饰的灿烂，顶饰的曲折，无一不体现了中国工艺的精湛。

Earth generates metal, Cloisonné, now classified as a national intangible cultural heritage, is a traditional Chinese metalcraft. The crafting of cloisonné makes use of bronze- and porcelain-making technology, and seamlessly blends traditional painting and carving techniques. Suffice it to say that the cloisonné assembles all illustrious characteristics of traditional Chinese craftsmanship. From gilded imperial palaces of the Forbidden City and University Hall of the Imperial Academy, to the Cloud Dispelling Hall of the Summer Palace, the glitter and shine of the ubiquitous cloisonné in formerly royal spaces is a timeless marvel. The Residence's two floors are all elegantly decorated with cloisonné objects – the luxury of the eight-treasured vessel, the serenity of the grand clock, the sparkle of chandeliers, and the exquisiteness of roof décor – all of which display the exquisiteness of Chinese art and craftsmanship.

金生水，水生木，木生火，火生土，土生金。五行相生相克，循环往复，生生不息。九层台中处处可见金镶玉的门把手，预示“金玉良缘”，完美地展现了“玉”与“木”的结合；楼梯建材选用的红酸枝和青色玉石进行结合，形成千年老树特有的木包石形态，隐喻“木石之好”；楼梯下放置墨玉荷花砚，楼梯上方设二十八星宿吊灯，表达了“水火相映”，更是巧夺天工，叹为观止。

九层台蕴含寓动于静、动静结合的哲学思想，张扬砥砺奋进、鱼龙变化的拼搏精神，追求上善若水、和谐平衡的智慧境界，将中华民族的文化精粹和核心价值浓缩到一起，成为陶冶性情的熔炉，传承非遗的学校，鉴赏艺术的殿堂，堪称立体形象的人生教科书。

Metal generates water; water generates wood; wood generates fire; fire generates earth; and earth generates metal. Mutually, cyclically, and eternally, the Five Elements reinforce and counteract each other. In the Residence, the ubiquitous gold-inlaid, jade doorknobs symbolize the harmonious union of gold and jade, as well as perfectly displaying the fusion of jade and wood. The staircases made of rosewood and blue jade revive the rare petrification of thousand-year-old trees, and metaphorically refer to the spirited goodwill between wood and stone. The dark-jade inkstone and lotus lying below the staircase, and the twenty-eight-constellation-themed chandeliers upstairs embody the contrast between water and fire. Both are products of unrivaled craftsmanship that surpasses nature.

Architecturally, Tai Residence gives expression to the philosophy advocating the unity of quiescence and activeness, and a dynamic quiescence, to the fighting spirit which emphasizes struggle and change, and to intellectual aspirations for water-like excellence and absolute harmony. It is actually a fusion of the essence and core values of Chinese culture. The Residencer plays multiple roles. It is a furnace refining human sentiments, a place of wisdom passing down intangible culture, a palace wherein fine art is enshrined, and a three-dimensional textbook on life.

卷尾语

ADDENDUM

文明守望，文化传承，文心雕龙，文以载道。九层台始于文化觉醒，立于文化品牌，重于文化传播。这里是交流空间，是展示平台，是居住场所，是艺术殿堂。欢迎大雅鸿达、墨客骚人，艺术精英，创业先锋，一同登高台。

Tai Residence is the watcher of civilization and the inheritor and embodiment of culture. It aspires to be the transmitter of *Dao*, or the Way. It is a brainchild of the Cultural Awakening; a cultural brand that is devoted to the dissemination of Chinese culture. The Residence is an space for exchange, a platform for exhibition, an ideal residence, and a palace of art. It is open to all who are renowned for their excellence, knowledge, artistic talents and business accumen. Let us gather together and ascend the Residence.

WHERE WE PROFOUNDLY AND INSIGHTFULLY REFLECT UPON THE COUNTRY, THE WORLD, THE TRUE SELF, AND THE ORIGINAL INTENT OF HUMANITY.

观宇宙
观天下
观自在
观本心

图书在版编目（CIP）数据

九层台 / 熊澄宇主编 . -- 北京：文化艺术出版社， 2018.1
ISBN 978-7-5039-6462-6

Ⅰ . ①九… Ⅱ . ①熊… Ⅲ . ①艺术摄影－中国－现代－摄影集
Ⅳ . ① J429.3

中国版本图书馆 CIP 数据核字 (2018) 第 014861 号

九层台

策　　划　梅岭创意经济研究院
主　　编　熊澄宇
创意摄影　贝德诺维基
撰　　稿　程　昱　山南居士
翻　　译　池　帧
审　　校　Thomas Garbarini
监　　制　九层台投资有限公司
责任编辑　董瑞丽　李　冬
装帧设计　北京雅昌设计中心
出版发行　文化艺术出版社
地　　址　北京市东城区东四八条 52 号　100700
网　　址　www.caaph.com
电子信箱　s@caaph.com
电　　话　（010）84057666（总编室）　84057667（办公室）
　　　　　84057691—84057699（发行部）
传　　真　（010）84057660（总编室）　84057670（办公室）
　　　　　84057690（发行部）
经　　销　新华书店
印　　刷　北京雅昌艺术印刷有限公司
版　　次　2018 年 1 月第 1 版　2018 年 1 月第 1 次印刷
开　　本　787 毫米 ×1092 毫米　1/12
印　　张　18
字　　数　图片 190 幅
书　　号　ISBN 978-7-5039-6462-6
定　　价　550.00 元

Concept: Meiling Institute of Creative Economy
Editor in Chief: Xiong Chengyu
Creative Photography: Beidenuoweiji
Copy Writing: Cheng Yu　Shannan Jushi
Translation: Chi Zhen
English Edrtins: Thomas Garbarini
Supervisior: Tai Residence Investment Limited company
Responsible editor: Dong Ruili　Li Dong
Format Design: Beijing Artron Art Design Center
Publishins: Culture and Art Publishing House
Address: No.52, DongSi 8 Alley, DongCheng District, Beijing, 100700
Website: www.caaph.com
E-mial: s@caaph.com
Telephone: (010) 84057666 (Chief editor room)　84057667 (Office)
　84057691—84057699 (Distribution Department)
Fax: (010) 84057660 (Chief editor room)　84057670 (Office)
　84057690 (Distribution Department)
Distribution: Xinhua Bookstore
Printing: Beijing Artron Art Printing Co. Ltd.
First Edition: January 2018　First Printed：January 2018
Folio: 787mm×1092mm　1 / 12
Seal: 18
Pictures: 190
ISBN 978-7-5039-6462-6
Price: 220 USD